YOUR KNOWLEDGE HAS VALUE

- We will publish your bachelor's and master's thesis, essays and papers

- Your own eBook and book - sold worldwide in all relevant shops

- Earn money with each sale

Upload your text at www.GRIN.com and publish for free

Eric Petermann

Mountain Hazards associated with Permafrost Degradation

GRIN Verlag

Bibliografische Information der Deutschen Nationalbibliothek:

Die Deutsche Bibliothek verzeichnet diese Publikation in der Deutschen National-
bibliografie; detaillierte bibliografische Daten sind im Internet über http://dnb.d-
nb.de/ abrufbar.

Imprint:

Copyright © 2008 GRIN Verlag GmbH
Druck und Bindung: Books on Demand GmbH, Norderstedt Germany
ISBN: 978-3-640-23950-4

GEOG 408 „Cryospheric Processes and Climatic Change"

Subjects: Geography, Geology
Date: 20th October 2008

Mountain Hazards associated with Permafrost Degradation

by

Eric Petermann

Outline

1. Introduction

Permafrost degradation due to global warming is a widespread phenomenon in recent decades (Lemke et al. 2007), latitudinal as well as altitudinal. Dealing with permafrost related mountain hazards requires knowledge on the exact permafrost distribution throughout mountain areas. Therefore, a monitoring project on European permafrost distribution will be discussed. Furthermore, it has to be determined, what mountain hazards exist in general and which of them could be affected by permafrost degradation in terms of frequency and/or intensity. Two different types of mountain hazards will be examined in detail: rockfall and outburst flood of permafrost dammed lakes.

Moreover, the impact of mountain hazards on the society will be marked. The focus will be on investigations which can be done, before undertaking engineering projects. After that a brief overview on perspectives in terms of climate scenarios and their possible impact on permafrost degradation and mountain hazards will be given. Finally, some concluding statements will be made.

2. Permafrost in mountain areas

The distribution of permafrost is the result of the interaction of a number of factors: Altitude, slope aspect, mean air temperature, solar radiation, snow cover, snow redistribution, wind and avalanches.

The occurrence of permafrost in middle latitudes is limited to mountain chains. Most research is done in Europe (Harris et al. 2001a; 2001b; 2003; 2005). The data density outside of Europe is markedly lower. Even the recent IPCC (Lemke et al. 2007) report concentrates mainly on data from European and some Asian mountain chains. A study by Li et al.(2008) revealed permafrost trends for China, which are similar to Europe.

The lower altitudinal limits of permafrost in Europe span a range from 1500 m in southern Norway to 2500 m in the southern Swiss Alps. Looking at the spatial distribution of permafrost, there is no continuous permafrost in mid-latitude Europe. This is caused by the steep terrain, the high solar radiation throughout summer and relatively high mean annual temperatures. Therefore, the spatial distribution of permafrost is discontinuous or sporadic. Another striking feature of permafrost in European mid-latitude mountains is that the permafrost temperature is just a few degrees below 0 °C. Hence, even slight climatic shifts can cause major changes in the depth of seasonal thawing and finally permafrost degradation.

The Permafrost and Climate in Europe (PACE) Project

The PACE (Harris et al. 2001b) consists of a transect of instrumented permafrost boreholes across the higher mountains of Europe (fig.1). The transect starts in Svalbard, includes the Scandinavian mountains, the Alps and ends in the Sierra Nevada in southern Spain. The aim of the project is to improve the knowledge on permafrost and, finally, the assessment of related hazards.

The project contains several complementary investigations like geophysical surveys, microclimatic investigations, numerical modeling of permafrost distribution and physical modeling of permafrost related slope

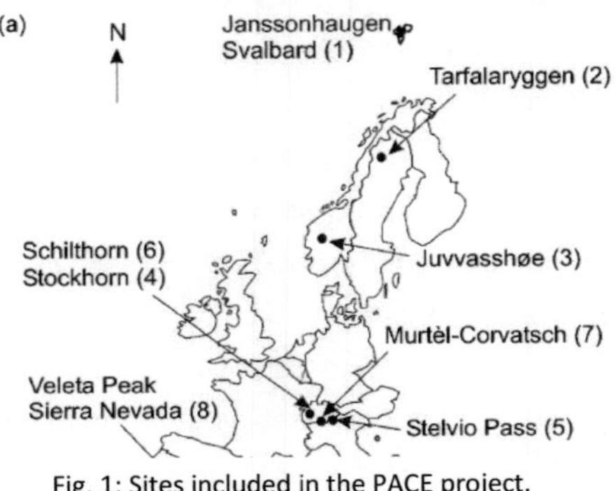

Fig. 1: Sites included in the PACE project. Harris et al. 2003

instability. The first results (Harris et al. 2003) showed that the thermal permafrost gradients are consistent with the 20th century surface warming. Important to notice is further the uniqueness of the thermal profile at each site (fig.2), which has a huge impact on risk assessment and related investigations required for determining the site specific features. For example, each thermal profile in fig. 2 indicates a site specific depth of the lowest temperature as well as different permafrost temperatures and active layer depths. Factors influencing the thermal profiles are geothermal heat flux (warming the permafrost at its bottom),

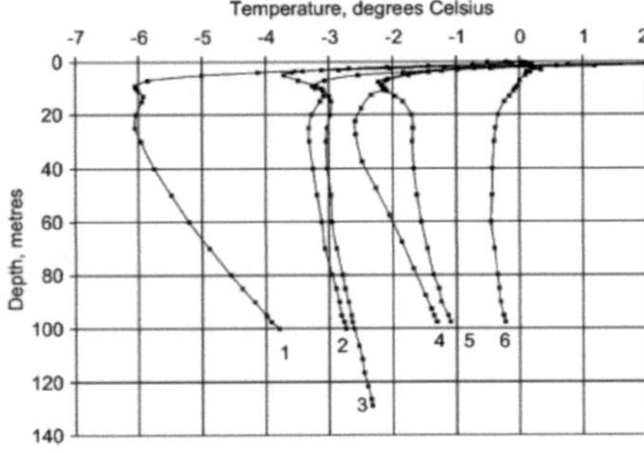

variations in lithology, the ground surface temperature and past changes of surface temperatures as well (Harris et al. 2003). Therefore, a big difficulty in predicting future permafrost changes is that the permafrost/thermal relationship is not in equilibrium, rather it is in part a function of the climate over the last decades and centuries (Harris et al. 2005).

3. Permafrost related hazards

Figure 3 summarizes some of the most common hazards occurring in mountain areas. This includes storms, volcanic eruptions, earthquakes, glacier advance, snow avalanches, lake outburst floods and mass movements. However, only two of them are directly related to permafrost degradation: mass movement and outburst floods.

Mass movements can be further subdivided into rockfall, debris flow, landslides and rock avalanches; outburst floods into floods resulting due to thermokarst or due to failure of mo-

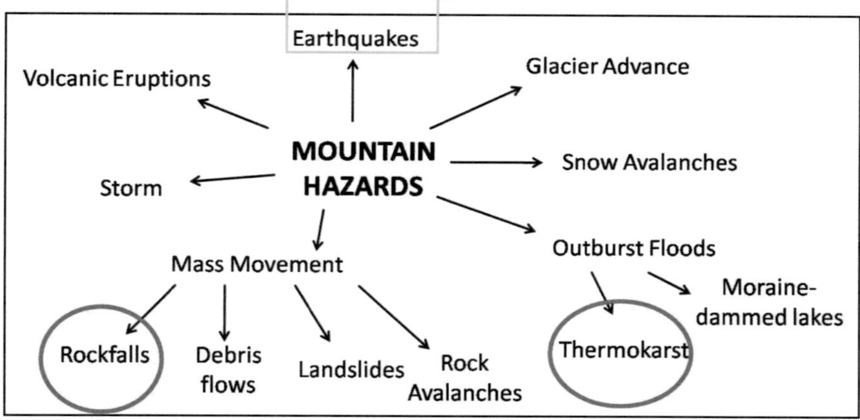

Fig. 3: Overview of common hazards occuring in mountian areas. Permafrost related hazards are further subdivided. Red encircled hazards wil be discussed in this paper later.
Own schema

raine-dammed lakes. An important point is that chain reactions of different hazards (e.g. rock fall in lakes causing floods) illustrate another issue, which has to be taken into account.

3.1 Mass movements/ Rockfall

Mass movements occur on different temporal and spatial scales. They span a range from solifluction and rock glacier movement with slow velocities to debris flows or rockfall marked by very high velocities.

The thawing reduces the strength of ice-rich sediment and bedrock. That leads to thaw consolidation in ice-rich soils. The slope failure ranges from shallow translational landslides in finer-grained sediment to rapid mudflows and debris flows (Harris 2005). Thereby, the critical issue is the ice-content of the frozen ground (Harris et al. 2001). Slope failure can even dam rivers. This results in the potential of floods as a consequence of sudden release of huge amounts of water in the course of dam failure.

The water percolation in highly fractured rock leads to thicker and earlier development of an active layer. Moreover, a convex topography (ridges, spurs, peaks) is subject to faster and deeper thaw than other areas. The melting of joint bonding ice causes 5 physical processes resulting in destabilization of the bedrock: Loss of bonding, ice segregation, volume expansion, increase in hydrostatic pressure, reduction of shear strength (Gruber &Häberli 2007).

A slope steepness of 37° is assumed as a threshold for separation of debris from bedrock slopes (Gruber & Häberli 2007). Warming surface temperatures causes active layer thickening, basal melting (leads to permafrost thinning) and hydrogeological changes in permafrost areas. For the release of a rockfall usually a specific trigger mechanism is required. Common triggers are earthquakes, high precipitation events and – on a different temporal scale – permafrost degradation. However, regarding past rock falls, the triggering mechanism is hard to reconstruct in terms of the exact timing, the initial topography and the rock properties.

Rockfalls are released due to rockwall failure and occur mainly on steep bedrock slopes with an inclination of at least 37°. Since, bedrock permafrost contains less ice than permafrost in sediments and the warming affects the bedrock at several sides in steep terrain, the degradation of permafrost in bedrock is rapid. Three surface features affect the character of the permafrost:

- Degree of fracturing (affect infiltration capacity)
- Snow and ice cover (affect rock temperature and water availability)
- Availability of water (affect advective transport of heat in fractured rock, rock weathering, turbulent exchange of latent energy at surface)

(Gruber & Häberli 2007)

A study by Fischer et al. (2006), analyzing rocktype in a rockfall area at Monte Rosa east face (Italy), illustrated a higher vulnerability of bedrock slopes in transition zones of different rocktypes; in this case para- and orthogneiss.

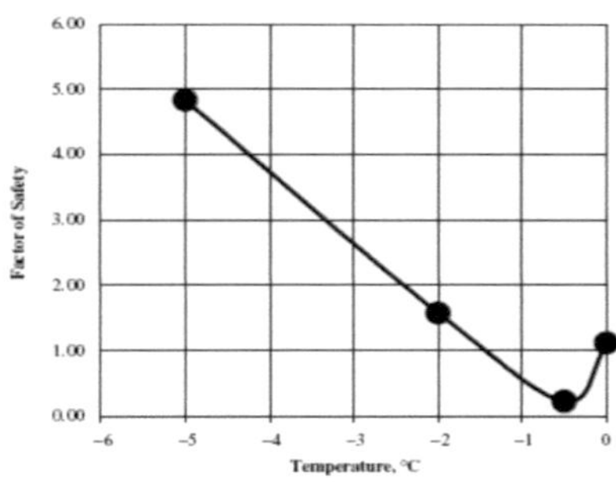

Fig. 4: Relationship between permafrost temperature and slope stability. The lowest factors of safety occur before thawing.
Davies et al. (2001)

Laboratory experiments (Davies et al. 2001) demonstrated the highest vulnerability of rock slopes at an inclination of 70° and a rock temperature of -0.5 °C. Consequently, the discontinuity is at temperatures of -1.5 °C to 0 °C less stable than in the thawed state (fig.4). This is caused by a reduction in shear strength, which leads to an increasing scale and frequency of slope failure.

A year with exceptional rockfall was the hot summer 2003 in Europe (fig. 5). It can be seen as an example for rapid destabilization as a consequence of extreme warming. The high frequency rockfall occurred during a summer that was ~ 3 °C (or 5 standard deviations) warmer than the average of the years 1960-1990, what caused thawing up to 4600 m. Moreover, more than 50 climbers were killed by rockfall (Schiermeier 2003).

Another fact that should be stressed is that no specific triggering effect like earthquakes or high precipitation events took place.

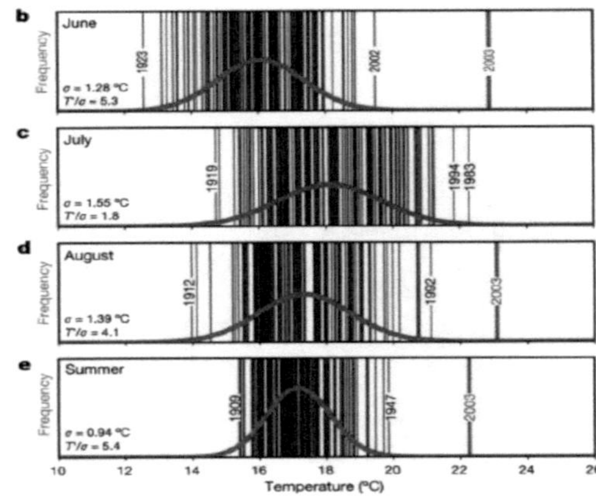

Fig. 5: June, July, August and summer mean temperatures for central Switzerland 1864-2003.
Schär et al. 2004

A study by Gruber et al. 2004 modeled rock temperatures for summer 2003 in Switzerland. The model is based on a monitoring program of rock temperature over a period of one year, starting in 2001. Data loggers in depths of 10 cm have been installed at 14 sites measuring the rock temperature every two hours. These 14 sites cover an altitudinal range from 2000-5000 m, all eight aspects and are installed at an average slope inclination of 70°. The

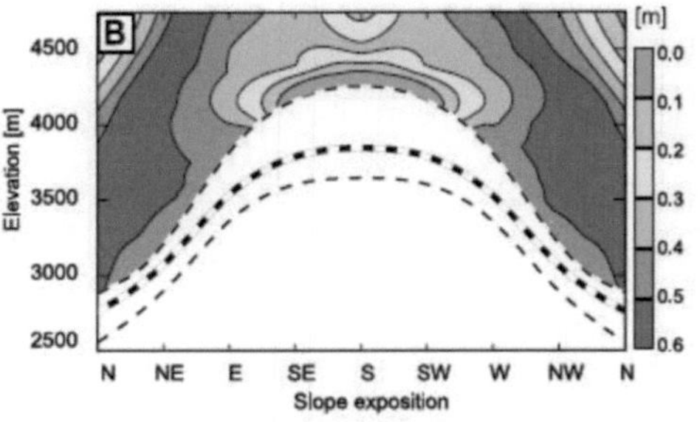

Fig. 6: Anomalies of thaw depth in 2003 compared to the maximum thawing 1982-2002 at different altitudes and aspects. The highest anomalies occurred on the northern slopes.
Gruber et al. 2004

data of those measurements were combined with climate data. Therefore, the surface energy balance was reconstructed, what was used to calibrate a model for reconstruction of rock temperatures using climate data as input.

In 2003 the active layer thawed 10-50 cm deeper than the maximum of the 21 previous years (fig. 6). As fig. 6 clearly shows the highest anomalies occurred on northern slopes. This can be explained by the fact that bedrock temperature on northern slopes is mainly driven by air temperature, whereas southern slopes derive a markedly amount of short wave solar radiation during summer as a second source of warming. Therefore, the dependence of northern slopes bedrock temperature on air temperature is much higher than those of southern slopes. During summer 2003 most of the rockfall events took place on northern slopes. Aside from higher anomalies of thawing depth; higher amounts of perennially frozen ground played a key role. Nevertheless, due to the low frequency of rockfall

Fig.7: Thermokarst lake on the Grubengletscher (CH).
Kääb &Häberli 2001

events it is difficult to relate one specific event statistically significant to permafrost degradation.

3.2 Outburst Floods

A good example for monitoring potential flood hazards and human intervention in terms of prevention are the thermokarst lakes on Grubengletscher in Switzerland. The development of one of these lakes will be discussed in detail. Thermokarst lakes usually occur in lowlands as a result of thawing of ice-rich permafrost or ice-rich bodies. In general, the highest concentration of such lakes outside the arctic area is in the Himalaya. Preconditions of lake growth are area wide permafrost caused sealing of the lake bottom, super-saturation of the ice-rich ground and a flat topography.

The thermokarst lake on Grubengletscher was photogrammetrically monitored since the 1960´s. The lake reached a final area of 10,000 m^2 and a volume of 50,000 m^3. Several outburst floods occurred in this area in the past. For example, floods with an outburst volume of 170,000 m^3 and eroding rates of 400,000 m^3 of debris damaged the village of Saas Balen in 1969 and 1970. The growth of this lake started in depressions of the rock glacier in the 1960´s with radial rates of 1.5 m/year (fig. 8). The growth accelerated during the

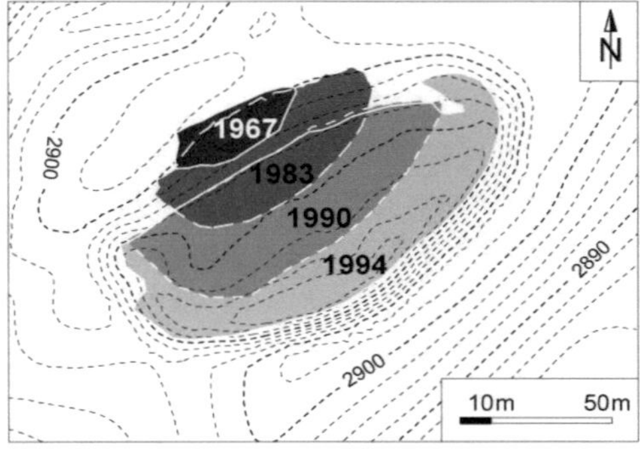

Fig. 8: Growth of a thermokarts lake on Grubengletscher in Switzerland. Radial rates accelerated during the 1980´s and 1990´s.
Kääb 2008

1980´s and 1990´s and became therefore more and more a threat. Consequently, the lake was artificially drained in 1995 (Häberli & Kääb, 2001, Kääb 2008). Hence, development of such lakes is typical for a warming climate, the likelihood of their occurrence in future will increase.

4. Risk assessment

Permafrost degradation even affects the stability of foundations in bedrock. Within the European Alps many of the infrastructure was built before the 1960´s when the permafrost "problem" was first recognized. Moreover, a lot of the infrastructure is situated in the critical lower permafrost limit (Gude & Barsch 2005). Therefore, to counteract, stabilizing geotechnical measures are recommended.

The methods of risk assessment figured out, refer to engineering project investigations proposed by Harris et al. (2001). The first phase contains a desk study with the aim of assessing the potential for and the character of permafrost-related hazards within an area. Therefore, different kinds of information are required: topography, permafrost distribution, slope element (gradient) and geological information (bedrock as well as superficial sediments). This data is combined using GIS techniques. Either these single attributes can be simply combined or weightings of these attributes can be determined and applied.

The second phase is aimed in validating and refining the hazard potential matrix. Therefore, thaw sensitive areas must be identified as areas where further ground investigations are necessary. For predicting the occurrence of permafrost an easy rule of thumb, solely requiring the bottom temperature of snow (BTS) during February and April, can be applied: exceeds the snow cover a thickness of 80 cm and the BTS is < -3°C the presence of permafrost is probable. If the BTS lies between -3°C and -2°C the presence of permafrost is possible and if the BTS exceeds -2°C no permafrost is present. For determining the character of frozen ground, geophysical methods can help to distinguish between ground in which water is unfrozen and largely frozen. Moreover it can be distinguished between ice-rich and ice-poor permafrost. Geophysical methods which are commonly used are: refraction seismic, ground penetrating radar, seismic tomography, DC resistivity tomography and electrical conductivity. Furthermore, borehole investigations are done.

5. Perspective

For this purpose, climate scenarios of the global average temperature are only of limited use. Rather, regional climate scenarios for mountain regions are needed. Using a regional climate model, Schär et al. (2007) propose a warming trend of 5°C until the end of the 21st century for Switzerland. Furthermore, they point out that the increased number of heatwaves (see 3.1) could be much more important than the rise in the annual average temperature.

Gruber et al. (2004) stated that climatic change does likely stronger effect mountain areas than the global average. But even in a warming world, regional cooling trends can occur. A reduction in snow cover and glacial ice cover may lead to greater winter cooling in certain areas (Wegmann et al. 1998).

Furthermore, if climate scenarios become reality it is likely that locations, frequency and magnitude of rock wall instabilities develop beyond the ranges of historic variability (Gruber et al. 2004).

6. Conclusion

Several studies clearly showed that permafrost degradation result in increased rockwall instability. The climate scenarios show a warming trend for upcoming decades, what would increase the likelihood of slope failure. Nevertheless, longer periods of monitoring are required to prove a statistically significant link between warming climate (e.g. permafrost degradation) and rockfall events.

A common way of evaluating the risk potential within a specific area can be done by a combination of GIS techniques and field surveys. Whereas, thermokarst lakes can be relative easily monitored by remote sensing, rockfall monitoring is much more complex due to the interaction of permafrost temperature, rock fracturing and other triggers apart from permafrost degradation. If a danger arises due accelerated growth of thermokarst lakes, they can be drained. Regarding bedrock slopes an increasing danger is not that obvious. Therefore, each risk assessment requires detailed knowledge of temperature-permafrost-stability relations and must be site-specific. The PACE project seems to be a useful approach in terms of incorporation of predicted changes in permafrost into practical guidelines accessible to engineers and land—use planners.

References

M.C.R. Davies, O. Hamza & C. Harris (2001): The Effect of Rise in Mean Annual Temperatures on the Stability of Rock Slopes Containing Ice-Filled Discontinuities. Permafrost and Periglacial Processes 12, 137-144.

L. Fischer, A. Kääb, C. Huggel & J. Noetzli (2006): Geology, glacier retreat and permafrost degradation as controlling factors of slope instabilities in a high-mountain rock wall: the Monte Rosa east face. Natural hazards and Earth System Sciences 6, 761-772.

M. Gude & D. Barsch (2005): Assessment of geomorphic hazards in connection with permafrost occurrence in the Zugspitze area (Bavarian Alps, Germany). Geomorphology 66, 85-93.

S. Gruber, M. Hoelzle & W. Haeberli (2004): Permafrost thaw and destabilization of Alpine rock walls in the hot summer of 2003. Geophysical Research Letters 31, L13504, doi:10.1029/2004GL020051.

S. Gruber & W. Haeberli (2007): Permafrost in steep bedrock slopes and its temperature-related destabilization following climate change. Journal of Geophysical Research 112, F02S18, doi: 10.1029/2006JF000547

W. Häberli, A. Kääb, D. Vonder Mühll & P. Teysseire (2001): Prevention of outburst floods from periglacial lakes at Grubengletscher, Valais, Swiss Alps. Journal of Glaciology 47(156), 111-123.

C. Harris, M.C.R. Davies & B. Etzelmüller (2001a): The Assessment of Potential Geotechnical Hazards Associated with Mountain Permafrost in a Warming Climate. Permafrost and Periglacial Processes 12, 145-156.

C.Harris, W. Haeberli, D. Vonder Mühll & L. King (2001b): Permafrost in the High Mountains of Europe: the PACE Project in its Global Context. Permafrost and Periglacial Processes 12, 3-11.

C. Harris, D. Vonder Mühll, K. Isaksen, W. Haeberli, J.L. Sollid, L.King, P. Holmlund, F. Damis, M. Guglielmin, D. Palacios (2003): Warming Permafrost in European Mountains. Global and Planetary Change 39, 215-225.

C. Harris (2005): Climate Change, Mountain Permafrost Degradation and Geotechnical Hazard. In: U.M. Huber (eds.): Global Change and Mountain Regions, 215 - 224.

A. Kääb & W. Häberli (2001): Evolution of a High-Mountain Thermokarst Lake in the Swiss Alps. Arctic, Antarctic and Alpine Research 33(4), 385-390.

A. Kääb (2008): Remote Sensing of Permafrost-related Problems and Hazards. Permafrost and Periglacial Problems 19, 107-136.

Lemke, P., J. Ren, R.B. Alley, I. Allison, J. Carrasco, G. Flato, Y. Fujii, G. Kaser, P. Mote, R.H. Thomas & T. Zhang (2007): Observations: Changes in Snow, Ice and Frozen Ground. In: *Climate Change 2007: The Physical Science Basis. Contribution of Working Group I to the Fourth Assessment Report of the Intergovernmental Panel on Climate Change* [Solomon, S., D. Qin, M. Manning, Z. Chen, M. Marquis,K.B. Averyt, M. Tignor and H.L. Miller (eds.)].

X. Li, G. Cheng, H. Jin, E. Kang, T. Che, R. Jin, L. Wu, Z. Nan, J. Wang & Y. Shen (2008): Cryospheric Change in China. Global and Planetary Change 62, 210-218.

Schär, C., P. L. Vidale, D. Lüthi, C. Frei, C. Häberli, M. Liniger & C. Appenzeller (2004): The role of increasing temperature variability in European summer heatwaves. Nature 427, 332–336.

Schiermeier, Q. (2003): Alpine thaw breaks ice over permafrost's role. Nature 424, 712.